This Book is Given With Love

To

From

Written and Illustrated by
Bobbie L. Lipman

ISBN 978-1-958032-14-5 (Hardcover Edition)
ISBN 978-1-958032-18-3 (Paperback Edition)

Registered in the Library of Congress.

Illustrations by Bobbie L. Lipman
Cover Illustration by Lindsay Lipschultz
Cover and Page Composition by Levi Stephen – 303 Pixels LLC (303pixels.com)

Printed in the United States of America.

Published by Here I Am Publishing, LLC.
info@hereiampublishingllc.com
Sandi Huddleston-Edwards, Publisher
780 Monterrosa Drive
Myrtle Beach, SC 29572

Acknowledgements

Creating a children's book and getting it published is a complex process. For me as a first-time author and illustrator, it took four years. Many times I put it away and thought that it would never be completed. However, due to the encouragement, knowledge, patience, and direction of the following individuals, it has finally come to fruition.

Many thanks to Daniel Klem, Jr., Ph.D., Professor at Muhlenberg College. Dr. Klem responded immediately to my request for information about birds and window strikes; provided vital resources from his fifty years of research; was genuinely supportive of the book project; and agreed to write the preface for *Can We Keep It?*. I also thank Dr. Christine Sheppard, Glass Collisions Program Director of abcbirds.org, for sharing resources from her very important work, including her manual entitled "Bird Friendly Building Design." Dustin Partridge, Ph.D, and Katherine Chen of New York City Bird Alliance are acknowledged for providing information about "Project Safe Flight" in New York City and bird-friendly building and window design.

I also thank my sister Jeanne for her ongoing support and assistance as my first editor and reviewer of the narrative and illustrations for *Can We Keep It?* As only a sister can do and as only an actual participant of a true story can do, she guided me with expert hands.

Many thanks are also extended to my husband and to our daughter, Amy Lipman, for cheering me on and not letting me give up on the book, especially the illustrations. To our son, George Hansen, I extend gratitude for connecting me with my editor and for assisting to review the project and scan it professionally. To Sandi Huddleston-Edwards, my editor and publisher, I am forever indebted. Your belief in me as an author and illustrator, plus your generosity, kindness, and knowledge, have been invaluable. Sandi and her colleague, Cynthia Valasin, led me and members of the Portside Writing Group through the process of expressive writing and book publication. I acknowledge each writing member's encouragement, talents, collaboration, and support.

To my granddaughter, Lindsay, I am grateful for your lovely painting that serves as the cover of *Can We Keep It?* You are a talented artist among all of your other amazing attributes.

I am a very blessed individual to have all of the above individuals in my corner.

Dedication

This book is dedicated to my dear sister, Jeanne Leader, who lived the story of *Can We Keep It?* She gave me wings to bring the cedar waxwing's story to life in words and illustrations.

This book is in loving memory of our dear mother, Jean Patricia Stull Leader, who also lived the story and originally shared it with me in her letter of March 1969. Her influence, love, creativity, and example as a Mom, learner, and lover of nature continues to guide me to enjoy each day.

Foreword

Birds fly into windows because they cannot see them, behaving as if windows are invisible to them. Windows are indiscriminate killers, taking the strong, as well as the weak. All bird species are vulnerable, but the ones that are killed most often are those found in or passing through the human built environment in urban, suburban, and rural landscapes. Scientists estimate that minimally 1.28 billion birds annually die from striking windows in the United States alone; that is an astonishing 3.5 million individual bird fatalities every day of the year. Billions more are killed worldwide. These deaths are always unintended and unwanted tragedies. Most, but not all, birds that hit windows succumb to their collision injuries.

This remarkable book, *Can We Keep It?,* is a true story of one bird that struck a window, was rescued, and lived. No matter how rare window collision survivors are, they are always celebrated, uplifting, and a source of great hope.

To ensure that all birds are protected from this lethal threat, we humans must make windows visible to all free flying birds so they can avoid them. Doing so will keep all caring and compassionate people like Jeanne and her mom from having to nurse any more birds back to good health. We know how to make windows in all human buildings safe for birds, and we should insist that this be done. Window-killed birds are always innocent victims that have no voice or other means to protect themselves. We humans must do that for them.

Please join like-minded people, like author Bobbie Lipman, who seek to save more bird lives from windows, as she has done in this emotionally moving story in this wonderful children's book. By joining Bobbie and others to make windows safe for birds, you will be helping to preserve bird life for us all, but especially for our children to discover the wonder and joy of these precious creations of Nature. No child should have to live in a world without birds.

Daniel Klem, Jr.
18 April 2024, Thursday

(Dr. Daniel Klem, Ph.D. D.Sc, Professor of Biology and Sarkis Acopian Professor of Ornithology and Conservation Biology at Muhlenberg College, Allentown, PA. He is the leading authority on birds and window strikes in the United States.)

Note from the Author: I am honored that Dr. Klem agreed to write this foreword.

It was a sunny winter afternoon. Snow sparkled on the treetops as we flew low, looking for dinner.

We spotted our favorite food: small blue berries on lots of juniper bushes. We began to swoop down.

I flew down fast to be the first one there!
Uh-oh! My head hit something hard!

I fell down to the snowy ground.
My flock didn't notice me. They were too busy eating the juniper berries.

After a while, my flock flew away.
For the first time ever, I was alone.

Suddenly I heard sounds. Something or someone was coming.
A young girl crouched down and looked at me.

I was afraid.
A girl named Jeanne
gently picked me up.
Her soft, fuzzy mittens
felt nice. I closed my eyes
as she carried me into
a warm place.

I heard her call, "Mom! Look!
I found a hurt bird. I think it flew
into the window. Can we keep it?"

I heard Mom reply, "Oh, Jeanne. It's a wild bird. I don't think we can keep it, but I will find out what to do."

Mom called a place named the Audubon Society.
They said to wrap me in a cloth to warm me and
to feed me berries and water when I was better.
They also said to let me go when my flock returned.

Mom brought something called a cage into the kitchen. She wrapped me in a towel. She put me in a box until I got better. Then I could go into the cage.

After a while I woke up. I saw Mom and Jeanne.
I seemed okay, so Mom put me in the cage. I hopped onto a perch and ate a berry. The next day I sat on Jeanne's hand.

I began to feel strong again. One day Jeanne let me out to sit on her hand. Suddenly, I started to fly! I flew all around the room. It was fun to fly again!

I heard Jeanne call to Mom, “The bird is flying all over!”
When I landed on a chair, Mom caught me with a towel.

After that, I had to stay in my cage. I missed my flock. Sometimes I rubbed my nose against the bars because I wanted to get out and fly.

One day I saw Jeanne rush into the family room.
She called out, "The flock is back!"

Quickly Mom and Jeanne
took me outside and
opened the cage door.
I called to my flock,
"See-sree! See-sree!"

They answered me: “See-sree! See-sree!” I was unsteady as I flew from the cage. I tried hard, and soon I reached my flock! After a while, we all flew away.

I told my flock all about Mom and Jeanne.
I was happy to be with my flock, but now I had to find
my own food, and it was cold outside.

One day we flew near Mom and Jeanne's house.
Maybe I can see them, I thought.
I flew off alone toward their backyard.

I landed on the fence by the juniper bushes.
I could see Mom and Jeanne through the kitchen window.

"See-sree! See-sree!" I called happily. I wanted them to know that I was okay. I wanted to tell them, "Thank you for taking care of me!"

Mom and Jeanne looked out at me.
I nodded my head up and down at them.
Then I called, "See-sree! See-sree!"

I was happy to see them smiling
and waving. After a while, I flew away.
I think they knew it was me.

How Can I Help? What Can I Do?

https://abcbirds.org Scroll down and click on "Glass and Birds". Click Menu to reach 3 tabs: "Prevention Information", "Recommended Products and Testing", "Programs and Resources". Information provided in each tab provides up-to-date resources for you to access by clicking. Under the last tab is a downloadable resource: Sheppard, Ph.D. 2021. "Preventing Bird Collisions with Glass: A Solutions Handbook".

https://allaboutbirds.org Enter into the search window: "Bird Collisions in Chicago". A list of articles will appear that describe how 1,000 birds perished by striking a convention center in Chicago in October, 2023, and how it can be prevented in the future.

https://www.audubon.org/news/reducing-collisions-glass Downloadable resources for preventing bird window strikes; access articles by clicking on links. Videos are also available.

https://www.birdmonitors.net Click on "Additional Resources" to find out how to help an injured bird or baby bird; how to minimize bird window strikes; research articles; info on bird migration and more.

https://birdsconnectsea.org Click on "About Us" at the top of the page. Click on "Our Work" to reach a series of tabs about what's happening in Seattle, WA to conserve birds. Click on the article "How to stop birds from hitting windows" to learn more information.

https://www.nycbirdalliance.org/our-work/conservation/project-safe-flight Project Safe Flight is a program in New York City that focuses on saving birds from window strikes.

https://nycbirdalliance.org/take-action/make-nyc-bird-friendly/make-your-windows-bird-friendly What you can do to make your windows safe for birds & prevent collisions.

https://www.nycbirdalliance.org/take-action/help-a-bird-in-trouble/what-to-do-if-you-find-an-injured-bird A detailed guide with instructions for actions to take when finding an injured bird. This guide is for New York City residents but it can be applied to any place that you live.

https://www.allaboutbirds.org/guide/search A general guide to enjoying and conserving birds.

Klem, Daniel, Jr., Ph.D. 2021. *Solid Air, Invisible Killer: Saving Billions of Bird from Windows*. A review of everything that is known about bird window collisions by the foremost authority on the subject in the U.S. and world. Hancock House Publications, B.C., Canada and Blaine, WA., USA. www.hancockhouse.com Also available on Amazon.

https://humanesociety.org Click on the magnifying glass at the top of the page. Type "Make your windows bird-safe" The article will tell you things you can do on the outside and inside of your existing windows; how to make windows bird-safe when you build or remodel plus how to help a bird who hits your window. Resources and products are also included.

About the Author

Bobbie L. Lipman grew up in Aurora, Colorado, in a busy household with four siblings and a variety of pets. Her mom taught her to draw and paint, to appreciate nature, and especially to enjoy birds.

Bobbie and her husband Jeff have four children. She was employed as a speech/language pathologist and public-school administrator for over thirty-five years. Bobbie retired in 2012 when she and Jeff moved to Myrtle Beach, South Carolina.

In March of 2020, while quarantined due to Covid-19 mandates, she discovered a letter from her mother that was postmarked March 12, 1969. It told the remarkable story of a cedar waxwing that collided with her parents' kitchen window and was brought indoors to recover.

Can We Keep It? tells the true story of the cedar waxwing. The character Jeanne is Bobbie's younger sister. The character Mom is their actual Mom. In the same month but fifty-four years later, Bobbie was visited by a flock of approximately fifty cedar waxwings. The three hundred photos she took of them served as models for the illustrations she created for *Can We Keep It?*

These days Bobbie and Jeff enjoy volunteering, exercising, traveling, gardening, reading, and spending time with friends, their four grown children, and their four granddaughters.

Milton Keynes UK
Ingram Content Group UK Ltd.
UKHW051951310824
447643UK00004B/15